Tropical Trees of Costa Rica

ISBN: 978-0-9705678-7-1

Printed in China

10 9 8 7 6 5 4 3 2

Developmental editor: David Featherstone
Book design: Ecce Gráfica; Zona Creativa, S.A.
Ecce Gráfica designer: Eugenia Picado
Zona Creativa designer: Gabriela Wattson

Published by Distribuidores de la Zona Tropical, S.A.
www.zonatropical.net

Tropical Trees
of Costa Rica

Willow Zuchowski
Photographs by Turid Forsyth

A Zona Tropical Publication

Introduction

The publisher was motivated to initiate work on this book—and its companion volume, *Tropical Blossoms of Costa Rica*—after witnessing the favorable reaction to the author's recently published *Tropical Plants of Costa Rica: A Guide to Native and Exotic Flora*. Comments from many readers suggested the importance of publishing a series of compact, inexpensively priced books for the general audience. Why, you might ask, add more titles to the existing stack of popular books about tropical plants? The publisher's response—bluntly—is that these books tend to be slapdash affairs. They consist for the most part of a rather uninspiring collection of photographs accompanied by brief textual notes about flowering times and geographic distribution—this in the best of cases.

Tropical Trees of Costa Rica thus attempts to combine striking photographs with engaging, textual nuggets, all written in a style that strives to make even technical notions clear to the general reader. Species descriptions treat a wide range of topics: plant-animal relationships; medicinal and psychotropic properties of various tree parts; importance of the plant to indigenous cultures—and to modern industrial society; use in horticulture; general geographic distribution (along with information about where the tree occurs in Costa Rica); and origin of common and scientific names. As you read through the book you'll see that it's not exclusively directed toward the general reader—many descriptions of plant parts rely necessarily on terms more familiar to the botanist than the general reader. In fact, although this book was written principally with a popular audience in mind, biologists not familiar with Costa Rican flora will find much of value here.

Species descriptions appear in alphabetical order—arranged by English common name. Consult the general index, page 93, to find a list of both scientific names and English and Spanish common names. If you know what a tree looks like—but you don't know its name—consult the visual index, beginning on page 87, to quickly find out if it is included in this book.

Note: Measurements are expressed in both metric and nonmetric terms. The metric measurements that are cited here come from the author's previous book, *Tropical Plants of Costa Rica*. When converting from centimeters to inches—and meters to feet—the editor rounded off numbers; in the case of converting from centimeters to inches, the resulting discrepancy is trifling.

Left: *Tabebuia ochracea* in bloom.

African tulip tree, Llama del bosque
Spathodea campanulata

Family: Bignoniaceae

The African tulip tree is immediately recognizable because of its colorful clusters of flowers. These consist of orange-to-scarlet, yellow-edged petals that form a scoop and then widen into five broad lobes. The showy flowering of this tree, which extends beyond the dry season, gives it its Spanish name, *llama del bosque* (flame of the forest). Another common name, fountain tree, refers to how water that collects in the calyx of an unopened bud shoots out, squirt-gun fashion, when the bud is squeezed. In addition to its striking flowers, the tree produces erect, woody capsules, the two halves of which are boat-shaped and contain many seeds. The African tulip tree is planted in many tropical areas; in Costa Rica, it is a widespread ornamental in parks and along streets.

Aguacatillo
Ocotea species

Family: Lauraceae

The common name *aguacatillo*, meaning "little avocado," applies to many species of trees in the Lauraceae family, including various species of *Cinnamomum*, *Nectandra*, *Persea,* and *Ocotea*. Of the 130 Lauraceae species in Costa Rica, about 75 of them occur in the Monteverde region. The majority of trees in this family produce small to medium fruits that are critical to the survival of the resplendent quetzal, whose altitudinal movement is determined largely by which species of Lauraceae are fruiting where. Quetzals, toucans, three-wattled bellbirds, and black guans disperse the seeds. One of these avocados, *Ocotea tonduzii*, is a common tree in the lower cloud forest. It grows to 100 ft (30 m) and produces small half-inch (1.5 cm) black fruits set in red cups. The photo above shows *Ocotea tenera*, a small tree that grows to 26 ft (8 m) tall; it occurs on both slopes, in a variety of wet forest sites around Costa Rica, including on the Pacific slope near Monteverde.

Akee, Akí
Blighia sapida

Family: Sapindaceae

Originally from West Africa and now planted in various tropical and subtropical regions, akee—sometimes called *seso vegetal* in Spanish—is most common in the Costa Rican Caribbean lowlands, although it is also grown occasionally in Guanacaste. This relative of the rambutan and *mamón* trees grows to a height of 35 to 65 ft (10 to 20 m). It has compound leaves with three to five pairs of glossy leaflets. The bulging, red (tinged with yellow) 3-lobed fruit has shiny, black seeds, each with an off-white aril. This aril, which looks like slightly spongy vinyl, is edible, but it is poisonous both before the capsule splits open and when it is overripe. The consequences of eating akee at the wrong stage are brutal and often fatal—nausea and vomiting are followed by a period of sleepiness and what appears to be recovery, but three to four hours later there is violent vomiting, then convulsions, coma, and death.

Angel's hair, Lorito
Cojoba costaricensis

Family: Fabaceae
Subfamily: Mimosoideae

The foliage, flowers, and fruit make this a decorative tree that could be used as an ornamental. The angel's hair tree—*lorito* or *cabello de angel* in Spanish—reaches a height of 50 ft (15 m). It has fernlike compound leaves and white, brushlike flowers in round heads. The fruit is a hanging red pod, up to 6 in (15 cm) long, with constrictions between the seeds. When the pods mature, they spiral open to reveal shiny black seeds. In Costa Rica, the tree is found in all major mountain ranges, including Monteverde. It also occurs in Panama. The similar *Cojoba arborea* can be seen along the highway at Palmares. The vivid red-and-black display of the twisted legumes on these trees appears to be a way of luring in seed dispersers, most likely birds, although they do not get many visitors.

Angel's trumpet, Reina de la noche
Brugmansia species

Family: Solanaceae

The angel's trumpet, or *reina de la noche*, gets its poetic names from its nodding or pendulous white or pink to peach flowers, which have very large, funnel-shaped to flaring corollas. The plant's intensely sweet fragrance is especially noticeable at night. Not only is the perfume strong, but the poisonous/hallucinogenic compounds in the plant are so powerful that ingesting it may induce temporary insanity. The various species of angel's trumpet originated in South America where indigenous groups use the plants medicinally (e.g., as a treatment for rheumatism) as well as in divination to commune with ancestors. In Costa Rica, several species and hybrids are grown as yard ornaments, usually at mid to high elevations. The trees are small—normally somewhere between 10 and 16 ft (3 and 5 m) tall.

Annatto, Achiote
Bixa orellana

Family: Bixaceae

Annatto is a popular dye plant that spans the cultures of industrialized countries and Amazonia. The yellow to red coloring obtained from the seed coating has been an important body paint, fabric coloring, and insect repellent to various native peoples of the neotropics, where it has been in cultivation for centuries. In the industrial world, bixin, present in the red-orange covering of annatto seeds, is a safe food coloring that not only lends color to cheese, butter, chocolate, and lipstick, but also provides a dose of vitamin A. It is also an ingredient in some oils, paints, and varnishes. In Costa Rica, where annatto is called *achiote*, it gives rice dishes a yellow-orange tint and imparts extra appeal to the cheddar, gouda, and edam cheeses of Monteverde. The shrubby tree may be seen in cultivation or, occasionally, growing wild. It has papery egg-shaped to heart-shaped leaves and delicately scented pink or white flowers that are 2 in (5 cm) across. The dull reddish-brown to bright scarlet fruit capsule is variable in size and shape, ranging from globose to ovoid, generally less than 2 in (5 cm) in diameter and covered with soft spines. It splits open to reveal about 50 small seeds with a red-orange coating.

Ant acacia, Cornizuelo
Acacia collinsii

Family: Fabaceae
Subfamily: Mimosoideae

This tree—also known as bull-horn acacia—appears from Mexico south to Colombia. In Costa Rica, it grows on the Pacific slope in dry areas up to 3,300 ft (1,000 m) and is common in the Guanacaste region. It is usually about 10 ft (3 m) tall, with tan-brown to copper thorns, around 1.5 in (4 cm) long, in the form of horns that are either straight or twisted. In addition to narrow leaflets, it has two or more glands on the base of the leaf stalk and a compact spike with minute bright yellow flowers. Its 1.5-in-long (4 cm) woody, brown pod contains a sticky yellow coating enveloping dark brown seeds. The ant acacia provides a classic example of ant-plant mutualism. Distinct bare areas on the ground, kept clean by the ants, surround the plants and protect them from climbing vines and competitors. When a potential herbivore brushes up against a tree, it is greeted by a frenzy of stinging ants emanating from the thorns at the leaf bases, and it wisely chooses to eat a different plant. In exchange for this protection service, the ants receive not only shelter from the conical thorns, but sugary secretions from the nectar glands on leaf stalks, as well as protein and lipids via food packets at the tips of the leaflets.

Avocado, Aguacate
Persea americana

Family: Lauraceae

The flesh of the avocado, with up to 30% oil, is so creamy-textured that vendors along streets in San José call out "*pura mantequilla*" (literally "pure butter"). The common name in Spanish, *aguacate*, is derived from an Aztec word that signifies *testicle tree*, probably a reference to the dangling fruits. Archaeological studies in Mexican caves indicate the avocado's use by people as early as 10,000 B.C. The avocado is rich in oleic acid, and research findings show antibiotic properties and steroids in the seed. The pear-shaped, oval, or globose fruit, sometimes more than 6 in (15 cm) long, is said to be an aphrodisiac. The fruit surface is smooth or bumpy, green to dark purple-black; it has yellow and green flesh, with a single large seed. Unripe fruit, seeds, bark, and leaves can poison livestock, fish, and other organisms. The tree grows to 100 ft (30 m), with a dense crown. The alternate, smooth-edged leaves, about 10 in (25 cm) long, are dark green above and gray-green below. They give off a distinct odor when crushed. The avocado tree—also called alligator pear or *palta*—has small, yellow-green, scented flowers.

Balsa tree, Balsa
Ochroma pyramidale

Family: Malvaceae

This pioneer species, a cousin of the kapok tree, grows in sunny open areas where there has been a fire or other disturbance. In Costa Rica, it is found on both the Pacific and the Caribbean slopes. The balsa tree is native to tropical America. It has a smooth, gray trunk and reaches a height of about 80 ft (25 m). Thick branches bear leaves with broad leaf blades—about 12 in (30 cm) long with a heart-shaped base—and bell-shaped, erect flowers about 5 in (12 cm) tall, with satiny-looking, cream petals folded back. These conspicuous flowers are nocturnal (bats visit them at night); they give off a scent like raw pumpkin.

The fruit capsule, which opens to reveal silvery-tan fluff surrounding many seeds, looks like a short, very fuzzy paint roller or microphone. The seeds, in their fluffy kapok coat, float through the air, but they also may be dispersed by water. In contrast to most tropical lumber species, which are sought out for their high density, balsa is valued for its lightweight wood that is ideal for rafts and model airplanes. Balsa wood was a component of airplanes during World War II. Native people of tropical America traditionally have used balsa in boat-making (*balsa* means raft in Spanish); it is now grown commercially in South America.

Brazilian firetree, Gallinazo
Schizolobium parahyba

Family: Fabaceae
Subfamily: Caesalpinioideae

The Brazilian firetree occurs from Mexico to Brazil; in Costa Rica, it is quite common on the road from Nicoya to Sámara on the Pacific slope. In Nosara, howler monkeys have been seen feeding on the large clusters of bright yellow flowers that adorn the branch tips. The adult tree has a smooth, light gray, straight, buttressed trunk and compound leaves. It grows to a height of about 100 ft (30 m). The 4-in-long (10 cm) fruit is flat, with only one seed. Young trees are very distinctive, with a nonbranching trunk and large, frondlike leaves that can be as long as 5 ft (1.5 m).

Breadfruit, Fruta de pan
Artocarpus altilis

Family: Moraceae

The breadfruit probably originated in New Guinea or the islands off Southeast Asia. It was introduced to the South Pacific Islands and later to the West Indies. In Costa Rica, it is seen throughout the lowlands, but is most popular on the Caribbean side. The tree, which reaches a height of 80 ft (25 m), has a milky latex. Its alternate, simple, pinnately lobed leaves grow to about 3 ft (1 m) long. The rounded fruit, 4 to 12 in (10 to 30 cm) across, is first green, then turns yellowish. Of the many varieties of the breadfruit tree, or *árbol de pan*, two occur in Costa Rica. One variety, called breadnut, has spiny fruits with seeds that, when boiled in salted water or roasted, taste somewhat like chestnuts. In Costa Rica, the common name for this variety is *castaña*, Spanish for chestnut. The other variety, whose fruits are smooth and seedless, is the typical breadfruit used in Polynesia as a breadlike vegetable (or as flour) when green, or as dessert when it is mature. The breadfruit tree played a pivotal role in the famous mutiny on the HMS *Bounty*. The crew's discontent on the return voyage from Tahiti partly stemmed from the fact that although fresh water was being rationed, it was given to breadfruit saplings bound for British colonies in the Caribbean. During the uprising, the small trees were thrown overboard.

Buttercup tree, Poroporo
Cochlospermum vitifolium

Family: Cochlospermaceae

Sometimes called the silk tree, this is a small deciduous tree—its maximum height is about 40 ft (12 m)—with smooth gray bark and lobed, tooth-edged leaves. The bowl shape of the large flowers makes the buttercup tree easy to distinguish from the many other yellow-flowered trees of the dry season. The capsular fruit opens to yield numerous small seeds, shaped like snail shells, that are embedded in white woolly hair; these fibers are good for stuffing pillows. The yellow-orange sap from the wood is used in dyeing cotton cloth. Large, pollen-seeking bees visit the flowers and vibrate the anthers to release pollen, some of which the bees transfer to other flowers. The buttercup tree grows from Mexico to northern South America; in Costa Rica, it is found at low elevations on the Pacific slope, generally in second growth. In other parts of the world, it is used as an ornamental.

Butterfly palm, Eureca
Dypsis lutescens

Family: Arecaceae

Originally from Madagascar and a few smaller African islands, the butterfly palm is found in Costa Rica as an ornamental from low to mid elevations, on both the Pacific and the Caribbean slopes. One of the most popular ornamental palms along city streets, it is also planted in yards and used indoors. It is sometimes called areca palm, bamboo palm, cane palm, or *palmera múltiple*. This species is often about 10 ft tall, and can reach 26 ft (8 m) or more, although the genus *Dypsis* includes some of the smallest palms in the world. The butterfly palm has clustered stems, and its trunk, which is yellow-green or gray-green, is ridged with old leaf scars. Although noted more for its arching foliage than its flowers, the palm produces a large, branching cluster of white flowers, followed by black-purple fruit.

Cacao
Theobroma cacao

Family: Malvaceae

The "food of the gods," as the genus name denotes, was frequently used in Aztec and Mayan ceremonies. Their version of hot chocolate was a mixture of cacao, hot chili pepper, and vanilla, with annatto for coloring and some cornmeal, perhaps added as a thickener. Cacao's origin seems to be in South America, but it is not clear how it reached Central America and Mexico, where the Mayans were the first to cultivate it. Since the pulp around the seeds is sweet and edible, humans on the move may have carried it as snack food, or possibly for medicinal uses, and dispersed the seeds during their travels. In Costa Rica, it grows in shade in the hot, wet lowlands of the Caribbean coast and parts of the southern Pacific coast. The tree reaches a height of 35 ft (10 m), but in cultivation it is often maintained shorter. The leaves are reddish when young, and the approximately 8-in-long (20 cm) ribbed fruit, with a bulging midsection, may be green, yellow, red-brown, or purplish. It contains seeds covered with white-to-pink, sweet, edible flesh. Researchers have found various components of chocolate that have stimulating, euphoric, and even hallucinogenic effects, but since these compounds occur in small amounts, it appears any "high" one gets from chocolate may be from the synergistic action of some of the hundreds of components.

Calabash, Jícaro
Crescentia cujete

Family: Bignoniaceae

When calabash fruits float down streams and into the ocean, they sometimes make their way to beaches as far away as Europe, via the Gulf Stream. The exact origin of *Crescentia* species is unclear, but they probably come from Central America and Mexico, or perhaps the West Indies. The calabash is a shrubby-looking tree of about 16 to 35 ft (5 to 10 m), with a crown of crooked branches and stubby shoots that end in clusters of simple leaves that narrow toward the base. Its flowers and fruits come directly off the trunk and branches. The flowers, which are pollinated by bats, are green-beige or yellowish, marked with reddish lines. The large globe-shaped fruits grow to 12 in (30 cm); the fruits of some trees are egg-shaped. The fruits have a woody rind and many small flat seeds in the pulp. In Costa Rica, the calabash is seen in yards and occasionally in pastures, especially in the Guanacaste region, to 3,900 ft (1,200 m). The tree is cultivated mainly as a source of calabash fruits for making decorative and utilitarian items such as cups and bowls.

Cannonball tree, Bala de cañón
Couroupita guianensis

Family: Lecythidaceae

The name of the cannonball tree comes from its large round fruit, 8 in (20 cm) in diameter, which contains smelly, fibrous pulp with a hundred or more seeds. The tree grows to 115 ft (35 m). It drops all of its leaves periodically during the course of a year. Clusters of flowers and fruits form on stalks that come directly off the tree trunk. The saucer-shaped flowers are fragrant, with six deep pink to red petals. Hundreds of pollen-bearing stamens form a ring in the center of the flower, while others line the hood that curves up and over the ring. It is difficult to trace the exact origin of this species—called *bala de cañón* in Spanish—since it has been planted as an ornamental for hundreds of years in many areas of Central and South America. The cannonball tree is a favorite of tropical botanical gardens because of its beautiful flowers and odd display of fruits. The many showy stamens make the flowers look like sea anemones. Visiting bees are attracted to, and collect, the sterile pollen in the hood section of the flower; while they are busy collecting, their backs get brushed with the true pollen from the center ring.

Cashew, Marañón
Anacardium occidentale

Family: Anacardiaceae

The cashew is thought to have originated in the region of Venezuela and Brazil, although its native range may be more extensive. It was introduced to Asia and Africa in the 1500s. In Costa Rica, it is planted on both slopes at low to mid elevations. It is a shrubby tree, usually less than 40 ft (12 m) tall, with simple leaves that are clustered toward the branch tips. New growth is reddish. The small flowers have a spicy cinnamon-like scent. The true fruit is the 1-in-long (3 cm) kidney-shaped gray knob, while what appears to be a fruit—the 4-in-long (10 cm) fleshy, yellow-orange-red part—is actually the fruit stem. Famous for its tasty nut, the cashew has many other uses. In Costa Rica, there is some small-scale processing of the nuts, but the cashew "apple," called *marañón*, is the main product. The tasty *marañón* is astringent and rich in vitamin C. Heat destroys the irritants in the fruit, so roasting a nut in its gray shell until it turns black makes it safe to peel and eat.

Cecropia, Guarumo
Cecropia species

Family: Cecropiaceae

Five species of *Cecropia* occur in Costa Rica, all recognizable from a distance by their candelabra-like branching and hand-shaped leaves. Since they grow well in disturbed areas and in light gaps, they are often seen along road cuts. Cecropias, sometimes called trumpet trees, are fast-growing; young trees may grow more than 12 ft (4 m) a year. Mature trees are usually less than 80 ft (25m) tall. The trunk is ringed with stipule scars and often has prop roots. The large leaves are deeply lobed with the leaf stem attached near the center. In many species, the hollow stems house *Azteca* ants, which attack and bite herbivores that might try to feed on the leaves. The ants get glycogen from the tree via tiny egglike packets called Müllerian bodies that form on feltlike pads near the base of the leaf stalks. Bats, birds, monkeys, and probably other arboreal mammals eat the fruits, which are small and packed densely in spikes. Three-toed sloths are easily (and therefore, frequently) spotted in the open crowns of these trees, where they go to sun themselves. The species shown here is *Cecropia obtusifolia*, which is widespread in Costa Rica and ranges from sea level to 4,750 ft (1,450 m).

Cerillo
Symphonia globulifera

Family: Clusiaceae

This understory tree is not very conspicuous except when in flower. When it is in bloom, the flowers attract butterflies and hummingbirds, as well as honeycreepers and bananaquits. The globose, bright pink buds bloom as white and pink-red flowers; the five petals curve inward and overlap in a way that makes the flowers resemble peppermint candies. *Cerillo* has bright yellow resin and green twigs with opposite leaves. The oval fruits have one or two seeds. The tree is found from southern Mexico to northern South America, as well as in Madagascar and Africa; in Costa Rica, it occurs in a wide range of elevations up to 6,000 ft (1,800 m). It usually grows to 50 ft (15 m) tall, but may reach 100 ft (30 m). It sometimes has stilt roots. In African samples of *cerillo*, researchers discovered the presence of the anti-HIV compound, guttiferone A.

Chicle
Manilkara chicle

Family: Sapotaceae

Hundreds of years ago, Mayan and Aztec people used a chewing gum quite different from the vinyl resins that we chew today. Their gum came from a sticky latex extracted from the trunks of chicle trees. The typical method of extracting the latex is by cutting diagonal slashes to intersect a vertical channel that carries the latex down the trunk, then collecting it in a container. The latex is later boiled down and formed into blocks. *Manilkara chicle* grows from southern Mexico to Colombia; in Costa Rica, it appears in moist to dry forests on the Pacific slope, up to 3,000 ft (900 m). The chicle tree—also known as *níspero* in Spanish—reaches a height of 100 ft (30 m) and has simple leaves, gray-green on top and yellowish green beneath. Its light brown fruit, which is up to 1.5 in (4 cm) long, usually contains two to five seeds, each just under an inch (2 cm) long with a scar that extends less than half the length of the seed. Another species, *Manilkara zapota*, has superior latex and is the better-known source of natural chewing gum, but it is less common in Costa Rica.

Fig tree, Higuerón
Ficus species

Family: Moraceae

For each of the 800 or so species of figs in the world, there is a matching species of pollinating wasp that appears to find its tree species by a chemical cue. These tiny insects pollinate flowers that develop on the inside wall of the syconium, a structure that looks like, and eventually becomes, the fig fruit. Species called strangler figs begin life as epiphytes in the crotch of another tree; they then produce roots that grow down to eventually anchor in the forest floor. As these roots grow, they expand in diameter, crisscross, and often fuse. A strangler fig eventually envelops its support tree, constricting and shading it. In figs, the conical twig tips are made up of two, often long, pointed stipules that cover new growth; these fall off, leaving circular scars. Guans, toucans, trogons, and other fruit-eating birds eat figs, as do bats, monkeys, and raccoon relatives. The size of the fruit varies from species to species, from half an inch to two inches (1 to 6 cm), and ranges in color from green to yellow to reddish. Approximately 50 species of Ficus occur in Costa Rica, ranging from sea level to high elevations, in wet and dry regions; some are planted for shade and ornament. Spanish common names given to various species are *chilamate, matapalo, higo,* and *higuito.*

Flamboyant, Flamboyán
Delonix regia

Family: Fabaceae
Subfamily: Caesalpinioideae

This tree has become popular throughout the tropics as an ornamental, not only for its magnificent blossoms, but also for its feathery foliage and shapely crown. It is actually rare in its native habitat of Madagascar, where an Austrian botanist first discovered it in 1828. For years, it was thought to be extinct, but was rediscovered in 1932. In Costa Rica, it appears as an ornamental in gardens and parks and along roadsides. The tree reaches a height of 35 to 50 ft (10 to 15 m), and it forms a dense umbrella-like crown of long leaves with numerous—sometimes up to 1,000—leaflets. The flamboyant, also known as the royal poinciana, has clusters of deep orange to scarlet flowers. Hummingbirds and butterflies visit the flowers of planted specimens, but the true pollinators in the wild are unknown. The showiest petal, also called a flag, folds vertically after the first day; but even old flowers, which fall apart over a series of days, remain colorful and probably help to attract insects and birds.

Frangipani, Juche
Plumeria rubra

Family: Apocynaceae

The frangipani is the national flower of Nicaragua, where it is called *sacuanjoche*. The tree—also called *cacalojoche* or *flor blanca*—is found from Mexico to Colombia and in the West Indies; it has been introduced as an ornamental in the Old and New World tropics. In Costa Rica, it is often found in rocky areas and is common along the Pacific coast. The flowers, which in Hawaii are favorites of lei makers, are seen year-round but are most profuse in April and May. They are pollinated by deceit—the white color and the perfume draw in nectar-seeking hawkmoths that carry pollen from one flower to another, but, upon probing the flower, are not rewarded with any nectar. The frangipani tree is usually less than 50 ft (15 m) tall. It has white latex and stout, tan-gray branches with conspicuous leaf scars, and long alternate leaves. In naturally growing trees, the flower clusters may have a few to dozens of sweetly scented flowers that are white with yellow in the throat; other color forms—purple, pink, yellow—exist in cultivation. The seed capsules form an upside down V.

Golden shower, Cañafístula
Cassia fistula

Family: Fabaceae
Subfamily: Caesalpinioideae

A golden shower tree in full bloom—its bright grapelike clusters of flowers dripping from the branches—is one of the most beautiful sights along roadsides in lowland Costa Rican towns. The hanging flower clusters, 20 in (50 cm) or longer, have many yellow-petaled flowers. The trunk of this medium-sized deciduous tree is cream-gray, and the fruit, a cylindrical dark brown pod up to 20 in (50 cm) long, has a scent of very ripe apples. Air pockets in each section of the seed capsule allow the pod to float on water surfaces, sometimes for as long as two years, before settling on a beach. This tree is one of a number of host plants for larvae of the large black witch moth. Originally from Asia, the golden shower tree is now cultivated in many tropical countries. In Nicaragua, the leaves are cooked and used as a laxative; its use may pose risks, however, since it can be toxic.

Guanacaste
Enterolobium cyclocarpum

Family: Fabaceae
Subfamily: Mimosoideae

This is one of the majestic tree species that grace the landscape of the Guanacaste region. It is found in lowlands all over Costa Rica, but it is most abundant on the Pacific slope. The tree's distribution ranges from Mexico to northern South America, as well as to the Antilles. Growing to a height of 115 ft (35 m), the spreading, low-branching guanacaste has a rounded crown in open areas. The large grayish trunk, more than 8 ft (2.5 m) in diameter, can be either smooth or have lenticels and fissures. The leaves, which are twice compound, fold at night. The small, white flowers appear in ball-like heads. The guanacaste—also called the ear tree—has shiny, dark reddish-brown pods that are ear-shaped, with compressed, twisted spirals. The tree attracts a wide variety of animals; small moths and beetles visit the flowers at night, and livestock and wildlife eat the nutritious fruits and seeds.

Guava, Guayaba
Psidium guajava

Family: Myrtaceae

In certain regions of Costa Rica, having a guava tree (or two or three) on your plot of land is not a choice—they just appear. And since they bear an edible fruit, are rather handsome trees, and provide firewood, people usually allow some to grow. Fruits that fall and ferment on the ground attract butterflies, including the beautiful morphos. To a novice, the many ripe, yellow guavas found in abundance under the trees are a real temptation, but these very ripe, tasty fruits are full of fly larvae (i.e., maggots). The exact origin of this tropical American native is unclear, since the tree is widespread in the neotropics and often associated with people. In Costa Rica, it is seen in most parts of the country. The guava is a small to medium-sized tree with a twisted, musclelike, smooth trunk; its branches have light bronze, green, and pinkish gray bark that chips off. The opposite leaves, which have minute translucent dots, are about 4 in (10 cm) long. White, lightly perfumed flowers appear in the leaf axils and have a brushlike center with many stamens. The popular sweet-smelling, yellow fruit is round (sometimes pear-shaped), with a dark scar on top and pink or white flesh inside. It has many small, hard seeds. Findings at archaeological sites in Peru and Mexico indicate that humans have been using guava for thousands of years. In Costa Rica, it is made into jam, jelly, paste, and filling for pastries and candies.

Gumbo limbo, Indio desnudo
Bursera simaruba

Family: Burseraceae

The gumbo limbo—also called the naked Indian tree, *indio pelado, jiñote,* and *caraña*—is found from Mexico to Colombia, in the Caribbean, and along coastal hammocks of Florida. It appears throughout Costa Rica, in dry to wet habitats, generally below 3,600 ft (1,100 m). This deciduous tree is among the most popular species used for living fence posts. It reaches a height of 80 ft (25 m) and has resinous sap. Its photosynthetic, greenish trunks are covered by copper-colored bark that peels off in thin, papery strips. When its leaves are crushed, they give off a pleasant pungent scent. The capsular fruit contains a single half-inch (1-cm), three-angled seed with thin scarlet flesh that is eaten by some birds. White-faced capuchin monkeys chew on new stem growth, a "pruning" process that may lead to a larger crop of fruit, which the monkeys, as well as peccaries, eat. Studies on gumbo limbo indicate that the dried bark has diuretic, cytotoxic, and antifungal properties.

Inga, Guabo
Inga species

Family: Fabaceae
Subfamily: Mimosoideae

Ingas, or *guabos* as they are known in Costa Rica, are common throughout the neotropics. There are around 300 species in the genus *Inga*, with more than 50 in Costa Rica, some of which are planted in coffee plantations because of their shade and nitrogen-fixing qualities. All species have once-compound leaves with an even number of leaflets and nectar-secreting glands between the leaflets. The leaf midvein is often winged, with leafy tissue spreading out on both sides. The flowers are brush-like, and the fruit contains seeds with a sweet, edible, white covering (aril). *Inga vera* (shown above), known as *guabo de río* in Spanish, is the most common member of this genus in the dry regions of Costa Rica, where it often grows near rivers. Bees, birds, and butterflies may visit its flowers during the afternoon; at night hawkmoths and bats appear. The type of sugar in its nectar changes at night, apparently through fermentation, from sucrose to the glucose and fructose that bats prefer.

Jacaranda
Jacaranda mimosifolia

Family: Bignoniaceae

The beautiful lavender flowers of the jacaranda attract attention from February to June. The clusters of bell-shaped flowers and fernlike foliage of this ornamental tree create a soft, cool impression that contrasts with the fiery intensity of orange, red, and yellow that many showy flowering trees exhibit. Originally from northwest Argentina but now planted all over the world for its shade and elegance, the jacaranda tolerates some cold weather. In Costa Rica, it grows at mid elevations. This deciduous tree has a smooth, gray trunk and reaches a height of 50 ft (15 m). Its disc-shaped fruit—which is about 2 in (6 cm) long, with a flattened, somewhat wavy edge—contains delicate winged seeds.

Kapok tree, Ceiba
Ceiba pentandra

Family: Malvaceae

Kapok trees are some of the most impressive trees in Costa Rica. This immense deciduous tree, with a straight trunk and a bulge or buttress at the base, has been revered by various indigenous cultures, including the Maya. Dugout canoes are still fashioned from the trunk, while doors and tables are made from the plank buttress. The tree is also renowned as the source of kapok used in life jackets and cushions, and for insulation. It grows to a height of 165 ft (50 m) or more, and to a width above the buttress of 6.5 to 10 ft (2 to 3 m). The crown has long horizontal branches and a somewhat flattened top. The kapok tree blooms when leafless, but does not bloom every year. Bats visit the white or pink flowers, arriving by the dozens after nightfall when the flowers open. The seed capsules split open to reveal hundreds of dark seeds in a cottony fluff (kapok). The tree is native to West Africa and to a region that spans from Mexico to Peru and Brazil; it is cultivated in Southeast Asia. Many children around the world have been introduced to this fascinating tree through Lynne Cherry's popular book, *The Great Kapok Tree*.

Mahogany, Caoba
Swietenia macrophylla

Family: Meliaceae

Of all the precious neotropical hardwoods, mahogany is perhaps the best-known and one of the most valuable. The beautiful high-quality wood, which is called *caoba* in Costa Rica, is made into furniture, cabinets, and musical instruments. The tree's gray and black/brown bark is vertically fissured into flat-topped strips and has a checkered appearance in older trees, which can grow to 130 ft (40 m). Mahogany has alternate, compound leaves and branching inflorescences made up of functionally unisexual, white to greenish, honey-scented flowers, a half-inch (1 cm) across. The distinctive capsular fruits, which take about a year to mature, are roughly 6 in (16 cm) long. When the erect woody capsule splits open, its five valves separate to release winged seeds. The 5-in-long (12 cm) seeds have a high germination rate, and the seedlings grow well in light gaps in the forest and in old pastures. A shootborer, the larva of the moth *Hypsipyla grandella*, often destroys the terminal growth of plantation trees, causing unwanted branching and thus lowering their commercial value.

Manchineel, Manzanillo
Hippomane mancinella

Family: Euphorbiaceae

The manchineel, a beach tree with a large, spreading crown, is known for its toxic properties and accidental poisonings. Eating the fruit—which resembles a small apple—results in swelling and pain from the mouth through the digestive tract; vomiting and diarrhea may be followed by shock, and even death if treatment, usually stomach-pumping, is not received. The 50-foot-tall (15 cm) tree has a grayish trunk that contains copious amounts of caustic latex, which indigenous people of the West Indies applied to their arrows. If you come into contact with any part of the tree, wash with sea water, or preferably soap and water. In parts of Florida, it has been eradicated, but it can still be found in southern Florida, in the West Indies, and from Mexico to northern South America. In Costa Rica, it appears only on the Pacific side, on the beach or close to it. The fruit is round, fleshy, and green (to yellow with a reddish tinge) with an odd-looking spiky pit that contains the seeds. The leaves have yellow-green to dark green, smooth, shiny leaf blades with a reddish gland at the base and small, gland-tipped teeth along the edge.

Mango
Mangifera indica

Family: Anacardiaceae

Native to parts of India, Burma, and China, the mango has been planted and naturalized in many tropical and subtropical parts of the world. There may be as many as 1,000 cultivars in India, where the fruit has a long history. In Costa Rica, the trees do best in warm areas with a distinct dry season. They provide shade and greenery in certain cities (e.g., in Alajuela's central park and on the road leading into Liberia). The tree, which reaches a height of 100 ft (30 m), has aromatic, resinous sap and a large crown of dense foliage. The alternate leaves are usually 10 in (25 cm) or longer, with new growth being reddish. Small flowers appear in large pyramid-shaped clusters at the branch tips. The fruit, which grows to 8 in (20 cm), is kidney-shaped or more rounded, or sometimes S-shaped. The mature fruit is usually a mixture of yellow, orange, and red. Besides being the basis for chutneys and sauces, the fruit, a good source of vitamin A, is delicious in ice cream, jams, and pies; it can be dried for later use. Mango trees contain mangiferin and isomangiferin, which have shown antiviral activity on the herpes simplex virus in laboratory tests. Although the resin is applied to various skin afflictions, it can be irritating, producing a reaction similar to that of the mango's cousin, poison ivy (Rhus radicans)—a rash with swelling, itching, and eye irritation.

Mayo
Vochysia guatemalensis

Family: Vochysiaceae

The *mayo* has a straight, grayish-white trunk and grows up to 100 ft (30 m). It has leaves that are opposite (or in whorls of three) and fragrant yellow flowers that form showy clusters. It appears from Mexico to Costa Rica, where it grows from the lowlands to about 3,300 ft (1,000 m) and also goes by the local names of *chancho* and *barbachele*. Research on the *mayo* in the La Selva/Sarapiquí area indicates that it has potential for use in reforestation. The shade of the young trees appears to suppress the growth of grass in old pastures that are being restored to forest; the grass often inhibits the germination and survival of tree seedlings.

These trees are also attractive perches for birds and bats that disperse seeds of forest tree species.

Monkey apple, Fruta de mono
Posoqueria latifolia

Family: Rubiaceae

The monkey apple, called *fruta* (or *guayaba*) *de mono* in Spanish, is a small tree that ranges from southern Mexico to South America. It grows in many parts of Costa Rica, especially in wet lowlands. Its 6-in-long (15 cm) narrow white blossoms emit an intense perfume at night, making it the quintessential hawkmoth flower. Monkeys and birds eat the 2-in (5 cm) fruit, which is yellow-to-orange and has angular seeds surrounded by pumpkinlike pulp. The bark has been used in infusions to treat diarrhea; a dusting of the dried flowers is supposed to repel fleas. A less common species, *Posoqueria grandiflora*, has flowers that may be twice as long as those of *P. latifolia*.

Monkey pot tree, Olla de mono
Lecythis ampla

Family: Lecythidaceae

Since this is a rare and endangered species, to encounter one of these trees—with its large, impressive, furrowed trunk, surrounded by fallen "monkey pots"—is a special event. The tree reaches a height of 130 ft (40 m) or more and has an impressive straight trunk with vertically fissured brown bark. Along with simple, alternate leaves, the monkey pot tree has flowers with six bluish or pinkish petals and hundreds of anthers. The large thick-walled, rounded fruit grows to 8 in (20 cm) in diameter. The seeds are edible, although some animals prefer the fleshy stalks, called funicles, that attach the seeds to the inner wall of the pot. Typically, the lid falls off while the giant downward-pointing fruit is still attached to a branch, making the seeds accessible to bats that are attracted by the funicles. These dispersers carry the seeds to a roost, eat the funicles, and drop the seeds. Squirrels and monkeys seek out the nutlike seeds. The tree grows in wet lowland forest from Nicaragua to South Amercica. Some people of northern Costa Rica use the seeds as an ingredient in candies, while the indigenous Kuna people of Panama use them medicinally.

Mountain immortelle, Poró
Erythrina poeppigiana

Family: Fabaceae
Subfamily: Faboideae

Native to Panama and South America, in Costa Rica this deciduous tree is planted in coffee and cacao plantations. The species belongs to a showy group of legumes known as coral trees, and the magnificent display of a flowering mountain immortelle is one of the highlights of the Central Valley in the dry season. The mountain immortelle reaches a height of 100 ft (30 m) and has prickly branches and trunk protuberances. The flowers have a pea-flower shape and are predominantly bright orange. Of the thirteen species of *Erythrina* in Costa Rica, many are small and are commonly used as living fence posts. They have decorative, red seeds containing poisonous alkaloids that are similar to curare arrow poisons in their effect on muscles. The tree is sometimes referred to as *poró extranjero* or *poró gigante* in Spanish.

Nance
Byrsonima crassifolia

Family: Malpighiaceae

The sweet-sour fruits of the nance tree are well-known by Costa Ricans. The fruit, which has a rotting-cheese odor, can be eaten raw; it is also made into preserves, candies, wine, and—at least in Brazil—ice cream. The flower petals are spoon-shaped, with a stem at the base and a wider "bowl" at the end. The squat, round, yellow fruits, each with a knobby cap, are a half-inch to one inch (1 to 2 cm) when mature. Female *Centris* bees collect the oil produced by glands on the flower calyx to feed to their larvae; they also incorporate it into the walls of underground nests. Birds eat the fruit. The nance—also called the shoemaker's tree or *nancite*—appears from Mexico south to Paraguay, as well as in the West Indies. In Costa Rica, it is typical of lowland Guanacaste's rocky savannahs, but it can be found at up to 4,600 ft (1,400 m) elevation. Red dye from the bark was formerly used for coloring leather, cloth, and wooden floors, while medicinal uses include treatment for diarrhea, chest colds, and fever.

Noni, Yema de huevo
Morinda citrifolia

Family: Rubiaceae

The natural range of this small shrubby tree appears to be Polynesia, Australia, India, and Malaysia. It was introduced to Hawaii and is cultivated in parts of tropical Asia; it is naturalized on the Caribbean shores of Costa Rica. The most striking aspect of this tree—the odd-looking, lumpy, green-white, and bad-smelling fruit—is sure to be noticed by Caribbean beachgoers. The ripe fruits, while not very appealing, are edible raw or cooked, as are the young leaves. The opposite leaves are large and simple, and the tree has green, conelike flower heads. Bees and ants visit around the flowers, perhaps scavenging nectar from old flowers. The tree is often 6 to 10 ft (2 to 3 m) but may reach 25 ft (8 m) tall. It is popular for its many medicinal uses, especially in Southeast Asia and the Pacific Islands. These include using the juice as a gargle for sore throat; taking a decoction of leaves and bark for tuberculosis; applying young fruit, with salt, on deep wounds; and using leaf poultices to relieve pain and reduce inflammation. Shampoo made with noni fruit is said to be insecticidal.

Oleander, Narciso
Nerium oleander

Family: Apocynaceae

This species, native to the Mediterranean and parts of Asia, has hundreds of cultivars, differing in flower color, foliage, and size. Oleander grows either as a shrub or as tree to 20 ft (6 m) or taller. Its leathery leaves are opposite or in whorls of three; the clusters of funnel-shaped, single or double flowers are white, pink, yellow, or red. The tree has long podlike fruits. This is a hardy ornamental capable of withstanding salt spray, wind, and other environmental stresses, including air pollution. All plant parts are highly toxic—oleandrin and other cardiac glycosides cause vomiting, diarrhea, abnormal heart beat and breathing, unconsciousness, and death. Despite this, it has been used as a home remedy for heart problems.

Orchid tree, Matrimonio
Bauhinia variegata

Family: Fabaceae
Subfamily: Caesalpinioideae

The orchid tree is common along the streets of San José. Originally from Asia, it is now found throughout the tropics as an ornamental. It is seen at low to mid elevations in Costa Rica, where it is sometimes called *palo de orquídea*. The tree usually is less than 20 ft (6 m) tall but may grow larger. Alternate, simple, two-lobed leaves with a cloven-hoof shape, about 4 in (10 cm) across, complement the orchid tree's purple-pink, 4-in-wide (10 cm) flower. The young pods are flat and may be suffused with red, turning to brown and gold as they mature. A full-grown pod measures about 8 in (20 cm) long. When mature, the fruits split lengthwise and remain on the tree; the spiral twists of the two sides of the pod suggest that the seeds are shot out explosively.

Peach palm, Pejibaye
Bactris gasipaes

Family: Arecaceae

The peach palm's origin, while unclear, is probably somewhere in South America, with migrating indigenous groups having introduced it throughout the region. Historically, the indigenous people of Costa Rica had peach-palm plantations. Today, it is cultivated from Honduras to northern Bolivia. The often-clustered trunks of this palm, which reaches a height of 65 ft (20 m), have wide bands of long, black spines. The one-seeded fruit is about 2 in (5 cm) long, growing in clusters of 200 or more. It is usually orange-red, but may be yellow or green. *Pejibaye* fruits—a good source of vitamin A, starch, and protein—are a common sight on street corners in San José, where vendors simmer them in metal boxes. Once peeled, they can be eaten as is, but because the nutty-starchy orange flesh is rather dry, a dab of mayonnaise is recommended. In some areas of the Caribbean slope, growing the tree for its heart of palm for local use and for export appears to be an economically promising activity.

Pentaclethra, Gavilán
Pentaclethra macroloba

Family: Fabaceae
Subfamily: Mimosoideae

This species of pentaclethra is found from Nicaragua to Panama and in certain parts of South America; in Costa Rica, it grows on the Caribbean slope to 1,650 ft (500 m). The tree does well in swamps as well as in better-drained areas. It tolerates poor soil and will grow in old pastures. Pentaclethra trees have broad crowns with compound leaves and reach a height of 130 ft (40 m). The erect, arching white spikes of flowers have a sweet, somewhat fermented scent; the two-valved, curved, woody brown pods reach a size of 12 in (30 cm). When the pods are mature, they split open explosively with a "pop" and eject seeds some 35 ft (10 m) away. Few animals eat the seeds, probably because of the presence of toxic compounds. Many seeds germinate, but between animal trampling and falling organic matter, about half of the seedlings perish.

Pink shower, Carao
Cassia grandis

Family: Fabaceae
Subfamily: Caesalpinioideae

In the middle of the dry season, the pink shower is one of the most conspicuous tree crowns of the Costa Rican dry forest. The distinctive large, pink flower clusters eventually turn orange-pink. The blossoms lack nectar and are buzz-pollinated by large bees seeking pollen. This deciduous tree appears from Mexico to northern South America, as well as in the Antilles. In Costa Rica, it is common in the Pacific lowlands, the Central Valley, and the Caribbean region (Limón and south). The brown-black, nearly cylindrical sausagelike pod, which grows to about 25 in (60 cm), is chambered, with each section containing a half-inch (1.5 cm) seed embedded in sweet pulp. This pulp is the main ingredient in *sandalada*, a Costa Rican drink made by cooking the pulp with milk. *Sandalada* is said to be a remedy for anemia and constipation. The tree is also called the coral shower tree or *sandal*.

Pink trumpet tree, Roble de sabana
Tabebuia rosea

Family: Bignoniaceae

The pink trumpet tree grows from Mexico to northern South America. In Costa Rica, it is found on both the Caribbean and Pacific slopes up to 3,900 ft (1,200 m). In the Central Valley, it can be seen, among other places, along Paseo Colón in San José. The tree provides shade in the wet season and displays a colorful crown in the dry season. The funnel-shaped flowers, in various shades of pink or white, have a yellow throat. Parts of this deciduous tree, including the twigs and leaves, are covered with small scales. The slender hanging fruit is capsular and contains many thin-winged seeds. Black spiny-tailed iguanas are known to eat the flowers. The tree's Spanish name, *roble* (oak) *de sabana*, denotes its high-quality wood, which resembles oak and is used in furniture and cabinets, as well as tool handles. Tea made from the tree's bark is a Costa Rican folk-medicine treatment for headache and colds. In the herbal-medicine world, *pau d'arco*, the common name given to some South American *Tabebuia* species, is known as a cure for a wide variety of ailments. Lapachol, which is present in some species, has antibiotic properties.

Pochote
Pachira quinata

Family: Malvaceae

Pochote is often seen as a living fence post in the dry areas of Costa Rica. This deciduous tree is found in dry to moist forest areas, from the central Pacific area north to Guanacaste. It generally grows from sea level to 1,650 ft (500 m), but sometimes as high as 3,000 ft (900 m). The trunk and branches of the *pochote* have thorns; it is sometimes called spiny cedar, or *cedro espinoso* in Spanish. In native forest, it can grow to 100 ft (30 m) tall, with a sizeable buttress and checkered, fissured, gray bark, with some large flakes. These large trees have distinctive horizontal and angular branching. The leaves are alternate and compound. The cream-colored flowers, which give off a delicate perfume at night, attract hawk-moths and bats. Appearing from Nicaragua to northern South America, the *pochote* is a source of lumber for construction, including doors and window frames; however, it contains a hygroscopic substance that will rust nails when the wood is used in humid conditions. Sometimes groups of silk moth larvae are seen on the tree trunks; large, colorful buprestid beetles lay eggs in the tree.

Powderpuff, Pompón
Calliandra haematocephala

Family: Fabaceae
Subfamily: Mimosoideae

The powderpuff (*pompón* or *pon pon rojo* in Spanish) is usually a large shrub, but it can reach 16 ft (5 m) in height. It has compound leaves and flowers in round heads about 3 in (7 cm) in diameter. The most conspicuous parts of the flower are the long red anther filaments, which attract hummingbirds. The fruits are flat, 2- to 4-in (6 to 10 cm) pods that split open when they mature. Originally from Bolivia, the powderpuff is common in and around Costa Rica's Central Valley, where it appears as an ornamental. Called *leuha haole* in Hawaii, it is cultivated for use in leis.

Provision tree, Jelinjoche
Pachira aquatica

Family: Malvaceae

This is one of the most conspicuous trees along the Tortuguero canals. It prefers wet places, growing at sea level and up to 3,600 ft (1,100 m). In addition to Costa Rica, the tree appears from southern Mexico to Peru and Brazil. It has compound leaves and grows to a height of 65 ft (20 m), sometimes with a buttressed trunk. What is noticeable about the provision tree (called *jelinjoche* or *poponjoche* in Spanish) are the large flowers and fruits. The phallic flower bud develops into five fleshy, straplike, tan or greenish white petals that reach a length of 12 in (30 cm). Bats pollinate the lilac-scented flowers. The rounded, reddish-brown fruit capsule is around 8 in (20 cm) or longer. The fruit and seeds disperse by floating. The provision tree makes a good ornamental both because its showy flowers and fruits can be seen during much of the year and because it may flower when it is only about 6 ft tall.

Queen's crape myrtle, Orgullo de la India
Lagerstroemia speciosa

Family: Lythraceae

Crinkly pink or purplish flower petals that resemble crape (a crinkled fabric) give this tree its common name. The queen's crape myrtle is a native of the Old World tropics, particularly China and India. In fact, it is also known as the pride of India, *orgullo de la India* in Spanish. In Costa Rica, it is a common ornamental tree around San José, Atenas, and Esparza. The tree is about 65 ft (20 m) tall with opposite leaves that are a lighter shade of green beneath. Various plant parts of this species are used medicinally; studies indicate that it has hypoglycemic activity and a poten-

tial use in treating diabetes. The durable wood of queen's crape myrtle makes it appropriate for boats and railroad ties.

Quick stick, Madero negro
Gliricidia sepium

Family: Fabaceae
Subfamily: Faboideae

Madero negro is sure to be seen by anyone traveling around Costa Rica since it is one of the most widely planted living-fence-post species. This small deciduous tree is usually around 35 ft (10 m) tall. In addition to compound leaves, it has fragrant 1-in-long (2 cm) flowers that are pink (sometimes more lilac or white) with a yellow center. *Centris* and *Xylocopa* bees pollinate the flowers, which form dense clusters along the branches. Pods resembling large flat green beans grow to be 6 in (15 cm) long. *Madero negro* appears from Mexico to northern South America and has been introduced to the Old World tropics. In Costa Rica, it is found on both the Caribbean and Pacific slopes, including the Central Valley, usually below 4,600 ft (1,400 m); it grows naturally in the dry forest of northwestern Costa Rica. The name *madre de cacao* (mother of cacao) is used in some cacao-producing regions where the trees are planted for their shade and nitrogen-fixing qualities. The protein-rich leaves also provide forage for cattle and nourish the soil as a green manure. Some Costa Rican farmers recommend making cuttings in May, at the time of the waning moon, and planting around the time of the first rains.

Quinine tree, Quina
Cinchona pubescens

Family: Rubiaceae

Of the several alkaloids from species in the Rubiaceae family that have played important roles in medicine, the most famous is quinine because of its capacity to prevent and combat malaria. A number of *Cinchona* species were exploited heavily in the 1800s as a source of natural quinine. The quinine tree depicted here is *Cinchona pubescens*, which ranges from Costa Rica south to Peru and Bolivia; it also occurs in Hawaii and the Galapagos Islands as an invasive species. In Costa Rica, this 16 to 65 ft (5 to 20 m) tree is found on both the Pacific and Caribbean slopes, from 2,600 to 5,900 ft (800 to 1,800 m). Its large leaves are more than a foot long (30 cm) and stand out as they turn red with age. Small pale flowers (pink or white) occur in large clusters at the branch tips. Wind disperses the tiny, winged seeds. There has been some speculation that the trees in Costa Rica are relicts from a time when the plant was cultivated by indigenous people. Scientists doing antimalarial research today are looking to wild quinine species as sources of valuable genetic material, because the microbes responsible for the disease are developing resistance to the synthetic medicines.

Rain tree, Cenízaro
Samanea saman

Family: Fabaceae
Subfamily: Mimosoideae

Also known as saman or *genízaro*, this tree is found from Mexico to South America and is grown as an ornamental in other tropical regions. In Costa Rica, it is most common in the Pacific lowlands, where it is one of the typical large—up to 100 ft (30 m)—rounded-crown species seen in Guanacaste pastures. It has very long, horizontal branches and fissured and checkered dark bark that cacti, orchids, and other epiphytes cling to. The leaves are compound and the 2-in-wide (5 cm) brushlike flower heads are pink and white. In the dry forest, it is a deciduous tree. The seed pods, which grow to 8 in (20 cm) long, contain a sticky pulp around their brown seeds. The "rain" of the name is really more of a mist produced from the excreted liquid of plant-sucking insects, such as cicadas, that feed in the canopy. The rain tree is a good choice as a shade tree because of its beautiful symmetrical crown. The wood makes attractive furniture and cabinets.

Red mangrove, Mangle
Rhizophora mangle

Family: Rhizophoraceae

Red mangrove is found where rivers and streams meet the sea, in the estuaries where fresh and saltwater mix and the tides cause fluctuations in the water level. The tree appears from Mexico to Peru, as well as in Florida, the Antilles, West Africa, and on islands of the Southwest Pacific. In Costa Rica, it grows in salty or brackish water on both coasts, along the waterway to Tortuguero, and often in river estuaries. Red mangrove varies in size, probably due to a combination of factors such as nutrients, salinity, and flooding. It can reach 80 ft (25 m), but is usually less than 40 ft (12 m). It has arching stilt roots and thick, opposite leaves. The fruit is about 1 in (2 to 3 cm) long, with one seed. This seed germinates while still on the branch, and an embryonic root that looks like a green bean with a brown tip extends down about 12 in (30 cm). The seedlings drop off after three to six months of growth. Sediments that settle around the roots of red mangrove result in land-building. Many insects and crustaceans depend on mangroves for food and shelter.

Ronrón
Astronium graveolens

Family: Anacardiaceae

This precious hardwood, which grows in secondary and primary forests, is a threatened species that is protected in a number of Costa Rican lowland sites, including Santa Rosa, Guanacaste, Palo Verde, Lomas Barbudal, and Carara. It grows from Mexico to Paraguay. The marbled wood, which is extremely heavy and hard, is suitable for bowls and other turned objects, cabinets, and flooring.

The 100 ft (30 m) or taller tree has a 3-ft-diameter (1 m) trunk that is light to dark gray and tan; the bark flakes off to create a patchy appearance. The sap has a pleasant turpentine-like odor. The tree's compound leaves turn red-orange in the dry season and are reminiscent of the autumn colors of sumac, its northern cousin. In addition to *ronrón*, it also goes by the names of *gonçalo alves* and *jobillo*.

Root beer plant, Hoja de estrella
Piper auritum

Family: Piperaceae

The presence of a chemical compound called safrole makes all parts of this plant smell like sassafras (anise). The leaves add flavor to meat and tamales. An infusion of the leaves may aid digestion, but be careful: studies of sassafras root bark showed that safrole produces liver cancer in rats. A fascinating use of the leaves is as fish bait in traps set by people living along rivers on the Caribbean slopes of Panama. The stems of the root beer plant have large leaf scars and prominent, sometimes-swollen nodes; the tree's leaves, covered with short, soft hairs, may be 20 in (50 cm) long, with an asymmetrical, heart-shaped base. The tree flowers and fruits year-round. The white flower spikes are thin and arching at first, then erect; they grow to 12 in (30 cm) or longer. Tiny fruits, which are relished by bats, are packed into thicker, darker, pendant spikes. Worldwide, there may be as many as 2,000 species of *Piper*, with around 124 representatives in Costa Rica; many of these are forest shrubs. Kava and black pepper are cultivated members of the Piperaceae family.

Rosewood, Cocobola
Dalbergia retusa

Family: Fabaceae
Subfamily: Faboideae

When the wood of this tree is cut or burned, it gives off a sweet fragrance that may have originally suggested the name *rosewood*. It is a medium-sized tree—reaching a height of 65 ft (20 m)—with a twisted dark trunk that has vertical fissures. It has compound leaves, clusters of white flowers, and flat legume fruits about 6 in (15 cm) long. The rosewood's very attractive streaked wood, which is known in Costa Rica as *cocobola*, is used in turned articles and cabinetry. Tropical hardwoods are more difficult to work than those of other, softer-grained trees, but their inherent resistance to insect damage and the color and grain patterns of the wood itself lend added value to the objects created from them. The tree is endangered due to habitat destruction and its extraction from natural forests.

Royal palm,
Palma real de Cuba
Roystonea regia

Family: Arecaceae

Like other palms, royal palms have served many uses in their native land. People eat the heart of palm and use the fruit as food for pigs; the leaves can be used for thatching. The smooth, light-gray trunk, which looks like a concrete pillar, bulges out about halfway up. It grows to a height of 80 ft (25 m), with a conspicuous smooth, green crown-shaft at the top. The compound, arching leaves are 10 to 13 ft (3 to 4m) long. The tree's inflorescence bud, which is long and erect, encloses not only the cream to white flowers, but also fluff made up of millions of tiny hairs that shower down to the ground beneath the palm. The genus *Roystonea* is named in honor of Roy Stone, a United States Army general. All of the world's ten species of royal palm are found in the Caribbean and neighboring countries. Various sources say the origin of *R. regia* is Cuba, but this species also appears to be native to adjacent islands, parts of Mexico's Yucatan coast, northern Central America, and southwest Florida. Elsewhere it is planted as an ornamental.

Sandbox tree, Jabillo
Hura crepitans

Family: Euphorbiaceae

This tree, found in the West Indies and from Nicaragua to South America, reaches a height of 115 ft (35 m) and a diameter of 5 ft (1.5 m). The sandbox tree is also called possum wood, dynamite tree, monkey's dinner bell, and, in Spanish, *jabillo*. It sometimes appears as an ornamental on city streets. The light tan to gray trunk has short, sharp, conical thorns and contains copious amounts of cloudy latex. The latex, which is caustic and irritating to the skin and can cause temporary blindness if it gets in one's eyes, is sometimes used by Latin American indigenous people to stun fish. The seeds are used in folk medicine as a purgative. Maroon male and female flowers appear on the same tree. The male flowers are arranged in cone-shaped clusters, while the larger, solitary female flowers have a starfish shape. The furrowed seed capsule—3 in (8 cm) across, with about fifteen sections—is shaped like a compressed pumpkin and is fairly heavy and woody. People fill them with hot lead to make paper weights and with sand to make ink blotters. The capsule is also explosive, splitting and shooting the disc-shaped seeds with great force a distance of at least 50 ft (15 m). The individual sections of the split capsule, reminiscent of dolphins, make interesting jewelry. Scarlet macaws feed on the seeds.

Saragundí
Senna reticulata

Family: Fabaceae
Subfamily: Caesalpinioideae

In Costa Rica, this small tree with fuzz-coated twigs is planted as an ornamental, but it grows naturally in the Caribbean lowlands as well as on the Pacific slope. The *saragundí* occurs from Mexico to Bolivia. It is identified by its compound leaves, yellow flowers, and by flat pods that are approximately 6 in (15 cm) long. The tree reaches a height of about 20 ft (6 m). The leaves fold up at night, a condition called sleep movement. If it is in fruit, the similar candle-stick senna or Christmas candle, which is also sometimes planted as an ornamental, is easy to distinguish from the *saragundí* because its pod has four longitudinal wings, in contrast with the *saragundí's* flat pod. The *saragundí* contains the compound rhein (cassic acid), which has antibiotic qualities. The plant has a wide range of medicinal uses, including preparations of leaves (and flowers) for liver and kidney ailments, venereal disease, snakebite, and rheumatism.

Tabebuia, Cortez amarillo
Tabebuia ochracea

Family: Bignoniaceae

During the dry season, about four days after a rain shower (or a sudden drop in temperature), whole populations of this yellow *Tabebuia* burst into flower. During this dazzling display, which lasts about four days, the trees are abuzz with pollinating bees. The tree—also called the yellow cortez—is about 65 ft (20 m) tall, with grayish bark that has shallow, vertical fissures. Tabebuia wood is heavy, hard, and long-lasting. The tree's opposite, long-stalked, compound leaves have five leaflets with a dense covering of star-shaped hairs on the underside. Dense, ball-like clusters of fragrant, yellow flowers decorate the trees; the legume-like seed pods, which are long, brown, and woolly, grow to 12 in (30 cm) long. The tabebuia is found in Central America and South America. In Costa Rica, it appears in dry to moist forest along the Pacific slope up to 3,300 ft (1,000 m), but is more common below 1,650 ft (500 m).

Tamarind, Tamarindo
Tamarindus indica

Family: Fabaceae
Subfamily: Caesalpinioideae

There is nothing like a refreshing *refresco de tamarindo* to quench your thirst on a hot day at the beach, but go light on it because it has a laxative effect. The drink is made by soaking the fruit, then rubbing the pulp off the seeds and adding sugar to offset the natural acids. It has high levels of vitamins C and B and also contains calcium and iron. This medium-sized tree grows to be about 65 ft (20 m) tall. It has compound leaves and a leathery reddish- to tan-brown fruit that bulges where the half-inch (1 cm) seeds form. Surrounding the seeds is an acidic, sticky, brown pulp. The tree's place of origin is most likely tropical Africa, but it is now found planted and escaped in the tropics and subtropics worldwide. In Costa Rica, it grows from sea level to 3,400 ft (1,000 m). Bagaces is the *tamarindo* capital of Costa Rica, with many large trees along the town's streets. In other parts of the world, tamarind is an ingredient in chutneys and sauces, including Worcestershire sauce.

Tonka bean, Almendro
Dipteryx panamensis

Family: Fabaceae
Subfamily: Faboideae

These majestic trees are crucial to the survival of an equally impressive bird—the great green macaw. Macaws—which feed on the tree's fruits and seeds, and nest in its cavities—appear to move to different regions between November and March in search of fruiting trees. Partially eaten seeds are able to germinate, depending on the extent of the damage. The *almendro*, as it is known in Spanish, is found from Nicaragua to Colombia. It is an emergent tree, growing taller than most other canopy species to become 165 ft (50 m) tall, with a smooth or somewhat flaky yellowish-white to brownish trunk. The trunk achieves a diameter of more than 3 ft (1 m) and sometimes has buttresses. The compound leaves have a straplike extension at the tip; the showy, purple-pink flowers occur in large clusters. The fruit of the *almendro* is about 2 in (5 to 6 cm) long and has several layers. In Costa Rica, the tree appears in the northeastern Caribbean lowlands. The lumber is attractive, heavy, and dense, and the tree is rapidly being turned into floorboards, threatening the existence of the great green macaw. Fortunately, *almendro* has become popular in native-species reforestation projects. The English name, tonka bean, usually refers to a South American species, *Dipteryx odorata*.

Tropical almond,
Almendro de playa
Terminalia catappa

Family: Combretaceae

This hardy seaside tree can tolerate high winds, salt spray, and sun, and it provides a welcome bit of shade at the beach. The Spanish common name *almendro de playa* means "almond of the beach." Originally from Malaysia and the Andaman Islands, off the coast of India (another name is Indian almond), the tree is now cultivated throughout the Old and New World tropics, including south Florida and the Keys. In Costa Rica, it is naturalized on the beaches of both coasts and also planted in parks. The bulging 2-in-long (6 cm) fruit is green to reddish yellow, edged all around with a keel. The white-to-reddish outer flesh is edible, usually acidic but sometimes sweet; the kernels in the very center have an almond or filbert taste, although this tree is not the source of the regular almond we eat. Ocean currents, as well as a variety of mammals, disperse the seeds, which are also relished by scarlet macaws. The tree may grow to more than 80 ft (25 m), but is often smaller. Its whorls of branches, in distinct tiers separated by branchless segments of the trunk, are especially noticeable in young trees. The clustered colorful leaves are yellow-green to red/orange/brown.

Weeping bottlebrush, Calistemón
Callistemon viminalis

Family: Myrtaceae

In the New World tropics, the weeping bottlebrush attracts hummingbirds and migrating Tennessee warblers; in its native Australia, it is visited by birds such as honeyeaters, lorikeets, and silvereyes. Bottlebrush trees are popular on city streets, both because they are able to tolerate a wide variety of environmental conditions and because they flower when young. The relatively small tree grows to about 35 ft (10 m). It has pendant branches and alternate, narrow leaves. Cylindrical inflorescences appear at the branch tips. The five tiny petals of the individual flowers are inconspicuous; however, the long, red anther filaments are very showy. The weeping bottlebrush flowers most of the year.

Wild cashew, Espavel
Anacardium excelsum

Family: Anacardiaceae

This large tree, which grows to a diameter of 6.5 ft (2 m) and a height of more than 130 ft (40 m), has a dark trunk and clusters of 12-in (30 cm) leaves that form a dense, dark crown. The local name *espavel* comes from *es para ver* (in order to see), which refers to indigenous people and explorers using this tree as a lookout point. The large clusters of green-white flowers, which turn to pink, have a clove scent. The kidney-shaped green fruit is suspended on a spiralling, fleshy stemlike receptacle. Bats and monkeys eat the receptacle of the fruit. While the toasted seeds are edible, the surrounding flesh has a caustic oil, cardol. The wild cashew belongs to the poison ivy family—other members include cashew, mango, pistachio, and the Brazilian pepper tree—and its resinous sap may cause a rash. Some indigenous people crush the bark and use it to stun fish in order to trap them more easily. These giant trees, with their dense foliage, are dominant features in many lowland *quebradas*, or stream beds. In Costa Rica, they are common on the Pacific side; they also appear in other Central American countries and south to Ecuador.

Wild nutmeg, Fruta dorada
Virola koschnyi

Family: Myristicaceae

The wild nutmeg appears in Central America and northern South America, reaching a height of 130 ft (40 m). The whorled, horizontal branching on the tree looks like spokes of a wheel when you look up into the crown. It has alternate, entire leaves with prominent venation, and its two-valved, orange-brown fruit, about 1 in (3 cm) in size, splits open to reveal a dark brown seed covered with a netlike red aril. The striking fruits attract medium and large frugivorous birds, which swallow the seeds and regurgitate them after removing the aril. Monkeys may also eat the arils. Resin from some species of *Virola* is famous for its use by Amazonian shamans in hallucinogenic snuff preparations and as an arrow poison. Various alkaloids in the resin cause a lack of muscular coordination, nasal discharge, visual distortion (seeing things larger than they are), nausea, and hallucinations.

Yellow elder, Vainillo
Tecoma stans

Family: Bignoniaceae

The yellow elder is a pioneer species that grows readily in rocky areas of the central and north Pacific slopes of Costa Rica. This relative of the trumpet creeper (*Campsis radicans*) goes by a number of names, including trumpet bush, yellow bells, *candelillo, sardinillo* (in Nicaragua), and *tronadora* (in Mexico). While the tree's form is not elegant, its attractive flowers give it ornamental value. These vase-shaped yellow flowers, which occur in clusters at the tips of the branches, give off a delicate vanilla or candylike scent. The leaves resemble those of the elderberry. The thin fruits are up to 10 in (25 cm) long. The tree's folk use in controlling diabetes has a scientific basis since the leaves contain certain alkaloids that have hypoglycemic effects. The plant also contains the antibiotic lapachol, as well as compounds that affect the liver.

Ylang-ylang, Ilang-ilang
Cananga odorata

Family: Annonaceae

This tree from southern India, Burma, Indonesia, and the Philippines is now cultivated in tropical areas worldwide, including Costa Rica, where it is an ornamental. It is found in the lowlands to an altitude of 660 ft (200 m). The tree grows to a height of 50 ft (15 m) and has long, pendant but upsweeping branches. The smooth-edged leaves are alternate and in a plane. The flowers, which are especially fragrant at night, are greenish or cream-yellow, with six petals. The flowers yield a scented oil, *ylang-ylang*, which is an ingredient of Channel No. 5 and other perfumes. In nature, the scent attracts beetle pollinators. The ylang-ylang tree is related to the soursop and custard apple.

Visual Index

Spathodea campanulata • 6
African tulip tree,
Llama del bosque

Ocotea species • 7
Aguacatillo

Blighia sapida • 8
Akee, Akí

Cojoba costaricensis • 9
Angel's hair, Lorito

Brugmansia species • 10
Angel's trumpet,
Reina de la noche

Bixa orellana • 12
Annatto, Achiote

Acacia collinsii • 13
Ant acacia, Cornizuelo

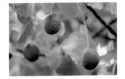

Persea americana • 14
Avocado, Aguacate

Ochroma pyramidale • 15
Balsa tree, Balsa

Schizolobium parahyba • 16
Brazilian firetree, Gallinazo

Artocarpus altilis • 17
Breadfruit, Fruta de pan

Cochlospermum vitifolium • 18
Buttercup tree, Poroporo

Dypsis lutescens • 19
Butterfly palm, Eureca

Theobroma cacao • 20
Cacao

Crescentia cujete • 21
Calabash, Jícaro

Couroupita guianensis • 22
Cannonball tree, Bala de cañón

Anacardium occidentale • 23
Cashew, Marañón

Cecropia species • 24
Cecropia, Guarumo

Symphonia globulifera • 26
Cerillo

Manilkara chicle • 27
Chicle

Ficus species • 28
Fig tree, Higuerón

Delonix regia • 29
Flamboyant, Flamboyán

Plumeria rubra • 30
Frangipani, Juche

Cassia fistula • 32
Golden shower, Cañafístula

Enterolobium cyclocarpum • 33
Guanacaste

Psidium guajava • 34
Guava, Guayaba

Bursera simaruba • 36
Gumbo limbo, Indio desnudo

Inga species • 38
Inga, Guabo

Jacaranda mimosifolia • 39
Jacaranda

Ceiba pentandra • 40
Kapok tree, Ceiba

Swietenia macrophylla • 41
Mahogany, Caoba

Hippomane mancinella • 42
Manchineel, Manzanillo

Mangifera indica • 44
Mango

Vochysia guatemalensis • 45
Mayo

Posoqueria latifolia • 46
Monkey apple, Fruta de mono

Lecythis ampla • 47
Monkey pot tree, Olla de mono

Erythrina poeppigiana • 49
Mountain immortelle, Poró

Byrsonima crassifolia • 50
Nance

Morinda citrifolia • 51
Noni, Yema de huevo

Nerium oleander • 52
Oleander, Narciso

Bauhinia variegata • 53
Orchid tree, Matrimonio

Bactris gasipaes • 54
Peach palm, Pejibaye

Pentaclethra macroloba • 56
Pentaclethra, Gavilán

Cassia grandis • 57
Pink shower, Carao

Tabebuia rosea • 59
Pink trumpet tree, Roble de sabana

Pachira quinata • 60
Pochote

Calliandra haematocephala • 62
Powderpuff, Pompón

Pachira aquatica • 63
Provision tree, Jelinjoche

Lagerstroemia speciosa • 64
Queen's crape myrtle, Orgullo de la India

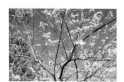
Gliricidia sepium • 65
Quick stick, Madero negro

Cinchona pubescens • 66
Quinine tree, Quina

Samanea saman • 67
Rain tree, Cenízaro

Rhizophora mangle • 68
Red mangrove, Mangle

Astronium graveolens • 71
Ronrón

Piper auritum • 72
Root beer plant, Hoja de estrella

Dalbergia retusa • 73
Rosewood, Cocobola

Roystonea regia • 74
Royal palm,
Palma real de Cuba

Hura crepitans • 75
Sandbox tree,
Jabillo

Senna reticulata • 76
Saragundi

Tabebuia ochracea • 77
Tabebuia, Cortez amarillo

Tamarindus indica • 78
Tamarind, Tamarindo

Dipteryx panamensis • 79
Tonka bean, Almendro

Terminalia catappa • 80
Tropical almond,
Almendro de playa

Callistemon viminalis • 81
Weeping bottlebrush,
Calistemón

Anacardium excelsum • 82
Wild cashew, Espavel

Virola koschnyi • 83
Wild nutmeg, Fruta dorada

Tecoma stans • 84
Yellow elder, Vainillo

Cananga odorata • 85
Ylang-ylang, Ilang-ilang

Index